Learning About ... Caves

Joy Ferguson

Copyright Information

Learning About ...
Caves

Joy M. Ferguson

ISBN 978-0-9877975-0-6

Public Relations Contact ~
butlermj2021@gmail.com

Dedication

With Love ~ For My Grandsons
Connor, Cooper, Everett, Grayson & Logan

TABLE OF CONTENTS

Beneath our feet is a hidden world, full of mysteries. When you think of a cave, do you imagine a strange, dark place, home to creepy critters, bears and bats – maybe a dragon or two? Or do you picture prehistoric peoples huddled around great fires, flames flickering off cave walls painted with ancient drawings of woolly mammoths? Or perhaps, you dream of a great and fun adventure, exploring the caverns and passageways in this sunken realm? Caves can be all of these things and more. They can tell us about prehistoric humans, climate change and give us clues to what life may be like on other planets! Let's start our journey into the incredible world of caves:

For the most fun with this book:

1. As you read the book, you will see words in **bold type**. These are names of famous caves and cave animals. To see incredible photographs of the places and creatures noted in **bold type**, simply use the Internet to search for pictures. You will be amazed! Why not try it now? Enter the words **Tham Lod Cave** into your search engine to marvel at images of this beautiful cavern in Thailand. Or, to see pictures of a cute animal that may sleep in a cave, enter the words **panda bear**. But beware! Caves are full of creepy critters. If you dare, search the name **African Cave Spider** for an Internet encounter with an alien-looking arachnid!

2. As you read the book, you will see words or groups of words written in *italic type*. You may not know the meaning of these. To learn the definitions for these difficult words, refer to the glossary on the back pages of this book. To try this now, go to the glossary and look up the word *speleologists* (pronounced: spee-lee-AWL-oh-jists) in the alphabetical listing there.

What is a cave?

A cave is often explained simply as a hollow space that has been created by nature and is large enough for a person to enter. *Speleology* (pronounced: spee-lee-AWL-oh-gee) is the scientific study of caves. The start of the word, “spele”, is related to the Greek and Latin words for “cave”; “ology” means, “the study of”. The person who studies caves is called a, *speleologist* (pronounced: spee-lee-AWL-oh-jist). *Speleologists* try to answer all sorts of questions about caves. How was it formed? What is it made of? What can the cave tell us about climate change? What types of plants and animals live in the cave?

Caves are found on every continent. A cave can be colossal: **Hang Son Doong**, in Vietnam, is about 4.8 kilometres long (over 3 miles) and, in places, almost 200 metres high (650 feet), meaning as tall as a 50-storey skyscraper! A cave may be deep: the **Veryovkina Cave**, located in Georgia’s Gagra Mountain Range plunges 2,212 metres (7,257 feet) underground – that is only about 400 metres (1,300 feet) less than the height of **Barbeau Peak**, the highest mountain in the Canadian Arctic! A cave can be ancient: **Jenolan Caves** in Australia are around 340 million years old, a time before even the mighty dinosaurs roamed the Earth! A cave may be one chamber or be made up of a series of tunnels and caverns: **Mammoth-Flint Ridge Cave System** in the United States contains about 644 kilometres (over 400 miles) of passageways – that is approximately the flying distance from London, England to Paris, France and back again!

Cave Zones

We often talk about three zones in the environment of a cave:

Entrance Zone

The **cave entrance zone** is where the cave opens to the surface world. It is usually a cool and shadowy space. Because it is so close to the surface of the earth, the temperature and amount of light in this area can vary depending on what is happening outside. The entrance zone of the cave usually receives enough light to be home to a variety of plants; many of the same plants that you would see just outside the mouth of the cave. If the entrance is bright, you may even find flowers, grasses and other plants. The entrance zone of the **Phraya Nakhon Cave** in Thailand has tall trees and lush plant life because it is open to the sun and rain.

Different parts of the cave system are home to different types of creatures. In the entrance zone, the surface world and the underground world meet. Because of this, you will find animals seeking shelter from the rain, sun or cold. Creatures in the entrance zone can be the same critters that you would find outside the cave. Two examples are **weasels** and rodents like the **Allegheny woodrat**.

Twilight Zone

Further inside the cave is the **cave twilight zone**. This is usually a cool and damp place that some daylight may reach, just enough to make it feel like dusk, just after the sun has set, but before the darkness of night. Here, you may find animals such as **cave crickets** and **cave beetles**, and plants like mosses and ferns. You may see fungi, like mushrooms; and bacteria that have formed mats on the cave walls. In the **Four Windows Cave** in the United States, bacterial mats – also called **lava tube wall slime** – are found in the twilight zone. In this cave, their colours range from silver to whitish-brown to gold.

Dark Zone

As you venture further into the cave, the light lessens until deep within you find the **cave dark zone**. This is an eerie, pitch-black world, filled with spooky, alien-looking creatures with bodies changed to adapt to the darkness.

Over time, often hundreds of years and many generations, creatures change or evolve to thrive in the dark environment. Nature may eliminate body parts that are of no use. Eyes are an example: in sunlight, vision is useful for hunting *prey* or escaping *predators*. But in total darkness, sight is of no use so these critters are blind. Another adaptation is the loss of colour or *pigment* so that creatures that live deep in the cave system are often white, pinkish or almost see-through (*translucent*). In the dark, surrounded by sightless creatures, colour serves no purpose. Eyes and *pigments* require energy for the body to create and maintain – nature doesn't waste energy that may be needed for other functions.

In return for losing some senses, such as sight, nature strengthens the senses that are useful to creatures that live in the dark, such as the sense of smell. Also, body functions may be adapted. For example, since food can be hard to find in the dark zone, creatures that live there are often able to survive with little food for long periods of time. And even parts of the body change, for instance the *antennae* (pronounced: an-ten-eye) of insects may become very long so that they are more efficient at smelling, feeling or sensing temperature, all of which would be useful in the dark, dangerous cave. Often these changes give the creature a monstrously bizarre or alien appearance. Search the Internet for images of **cave springtails** and **cave isopods**.

Types of Caves

Nature can shape caves using the elements: water, earth, fire and air.

Here are some examples of how caves are formed:

Solution Caves

Most of the world's caves are solution caves. This means they were formed by the chemical and physical actions of water and earth. Solution cave formation begins with rainwater: when it rains, water trickles along the ground, some of it filling creeks, ponds and lakes. Some of it soaks into the soil, giving thirsty plants a drink. But there is also rainwater that seeps deeper into the earth. As the water travels down, it picks up the gas from the soil. In this example, the gas is carbon dioxide. This turns the water into a weak *acid*, called carbonic acid – *acids* are chemical substances that can dissolve or eat away some other materials. This acidic water is able to dissolve bits of limestone and some other types of rock as it passes through them. The water may also encounter other harder rock that it cannot break down and so will be directed back to the softer limestone. This eating away of limestone creates the twists and turns in the passageways of the cave system. Over hundreds, thousands and millions of years, the acidic water flows, enlarging existing cracks and creating new ones. Bits of sand and stone are picked up by the flowing water and also help to wear away the softer rock until tunnels and caverns are formed. This "wearing away" is called *erosion*.

On the Internet, you can see wondrous photos of limestone caves around the world, for example: Canada's **Castleguard Cave**, **Cave of the Mounds** in the United States, Peru's **Gruta de Huagapo**, **Gouffre Berger** in France, **Jeito Grotto** in Lebanon and Japan's **Ryusendo Cave**, to name only six.

Sea Caves

Waves crashing against the bottom of cliffs on the shores of oceans, seas and lakes can wash away weaker types of rock. The water rushes with immense force, hitting the cliffs and gauging the soft rock. The waves also carry sand and pebbles that beat against the rock-face, helping with *erosion*. The wave action can create an opening that may widen and deepen over time to become a large sea cave.

Fingal's Cave on Staffa, an island off the Atlantic Ocean coast of Scotland, is made of basalt, a volcanic rock formed when *lava* cools rapidly. Fifty-nine million years ago, **volcanic eruptions** created the island and formed spectacular six-sided columns of basalt. Cracks and lines of weakness in the basalt allowed wave action to slowly carve Fingal's Cave.

This sea cave is famous for the eerie echoes made by the waves: the arched cathedral-like roof and basalt-pillared walls echo the music of the rushing water. In 1830, after visiting the cave and listening to the magical sounds, famous composer Felix Mendelssohn wrote "The Hebrides Overture", sometimes called "Fingal's Cave". In the Gaelic language widely spoken in ancient Scotland, Fingal's Cave was fondly called "The Cave of Melody". Internet videos let you hear both the overture and the actual sounds made inside Fingal's Cave.

The **Blue Grotto** is located on the island of Capri in Italy. This true sea cave is famous for its beauty and as the cavern where the Roman emperor Tiberius lazed in its waters. To add to the mystery of this sea cave, you can only enter at certain times of the year and when the tide is low. You must lay flat in a small boat to slip through the low opening. Once inside this cave, the sea water sparkles a brilliant sapphire blue.

Not all underwater caves are true sea caves. Instead of being created by waves, some have been made in other ways. The Earth's sea levels have changed over thousands and millions of years. What is now an underwater cave may once have been on land, having formed as a solution cave. But then, as the sea's water level rose, it filled the solution cave, making it an underwater cave, but not a true sea cave. The stunning **Caves of Phang Nga Bay** in Thailand are limestone solution caves that the sea has claimed.

Glacier Caves

A *glacier* is formed when snow falls and is packed down over time, perhaps even many thousands of years, to become a thick frozen mass. Gravity and the weight of a *glacier* move it forward. These huge sheets are made of snow, ice, rock, sediment and meltwater. Over time, melted water excavates tunnels through the *glacier*, creating caves. Similar to the rings in a tree trunk, layers of ice in *glacier* caves hold information about the Earth's weather conditions in the years they formed, telling us about climate change over time.

Iceland's **Crystal Ice Cave** is a stunning *glacier* cave. Like most *glacier* caves it is blue. In particular, this cave is famous for its beautiful turquoise ceiling. Since 2011, this cave has formed each summer when a large glacial river of meltwater has flowed.

In Washington State in the United States, after the **volcanic eruption of Mount St. Helens** in 1980, a *glacier* began to form. From fire has come ice. Fiery volcanos often form hills or mountains as rock and ash build up during repeated eruptions. If the mountain is high enough or located in a cold climate, ice sheets may form. The **Caves of**

Crater Glacier on Mount St. Helens are true *glacier* caves. This *glacier* is thought to be the world's youngest and it is still growing – now that is cool!

Glacier caves are found in many countries. In Norway, the **Nigardsbreen Glacier Cave** was discovered to have about an 8 metre high dome (26-foot), and is 30 metres long (98 feet) and 20 metres (65 feet) wide. This magnificent ice cathedral shimmers with magical shades of blue ranging from a light turquoise to a deep sapphire to violet.

Blue *glacier* caves are only found deep inside ice sheets. Why are they blue? Isn't ice white? The frozen water or ice we are used to seeing is white because when the water freezes it traps air bubbles – think of the white centre of an ice cube. But, large amounts of water appear blue – think of the blue of the ocean. In *glacier* caves, the frozen water appears as blue as the ocean because when ice crystals in a *glacier* are compressed by its weight, the air inside bubbles out. As this happens, the ice crystals become larger and denser. Without the air inside making them white, these large sheets of ice crystals – large amounts of frozen water – appear the jewel-like blue of an ocean and sparkle like blue diamonds. Depending on what minerals and *organic matter* are also trapped in the ice, there will be variations in colour. For instance, if algae are trapped in the ice, the blue may be even brighter, and in a few *glacier* caves, the ice takes on a purple tint.

Ice Caves

The term "ice cave" does not have the same meaning as "*glacier* cave". An ice cave first forms as other caves, such as a solution cave. Once the solution cave exists, it can be called an ice cave if it stays so cold that there is ice in the cave all year round. Ice caves hide sparkling secrets, beautiful ice formations that glitter in lantern-light. You may see glazed over lakes, frozen waterfalls and frosty feather-like ice crystals.

An ice cave – not a *glacier* cave – in Austria called the **Eisriesenwel** – German for "World of the Ice Giants" – is a chilly cave system of caverns and tunnels that reach 42 kilometres (26 miles) into the earth. Over thousands of years, a river wore away sections of rock in the mountain creating the labyrinth of passageways. Tourists are allowed to visit about the first kilometre (half-mile) of the Eisriesenwel cave network. Nature's dazzling ice formations and icicle sculptures glisten in the World of the Ice Giants.

Lava Caves

Magma is hot liquid and semi-liquid rock boiling below the Earth's surface. When this extremely hot *molten* (pronounced: mole-ten) rock erupts from a volcano, it is called *lava.*

If you have ever shaken a bottle of cola before opening it, you have seen a soda pop eruption! The cola expands violently, shooting out of the bottle and into the air. A **volcanic eruption** is even more explosive! During a **volcanic eruption**, the *molten* rock, once trapped underground, shoots through a small opening or vent in the earth.

The fiery *lava* can form a type of cave called a lava tube: as the *lava* flows down the slope, spreading out from the volcano, its surface comes into contact with cooler air. This colder air causes the outer portion of the lava to cool rapidly, forming a solid crust, much like ice forming on the surface of a river. But beneath that hard surface, extremely hot liquid *lava* continues to surge. Great rivers of this *molten* rock begin to carve trenches into the earth, similar to how a river of water scrapes a channel out of the land. When the trench dug by the *lava* is surrounded by cooled *lava* that has hardened into rock, a tube of solid volcanic rock with a fast-flowing river of *lava* running through it has been formed.

When the volcano stops erupting, these underground rivers of *molten* rock flowing through the rock tube can break through the end of the cooling crust to escape. Now, without *lava* flowing through them, they are caves. This tunnel-like cave is called a *lava* tube.

The cooling of flowing and dripping *lava* can sculpt beautiful shapes, some similar in appearance to those found in limestone caves. Because Hawaii was formed by volcanoes, it has countless *lava* tubes beneath its surface. Hawaii's **Kazumura Cave** may be the world's deepest and longest *lava* tube system at more than 1,100 metres deep (3,600 feet), 65 kilometres long (40 miles) and, in sections, opens to 18 metres (60 feet) in height and 21 metres (70 feet) in width.

Wind Caves

Wind caves, also called Aeolian (pronounced: ee-OH-lee-un) caves – named after Aeolus, the Greek god of wind – are less common than most other types of caves. They are usually found in desert cliffs. Aeolian caves are formed by gusting winds carrying sand and bits of stone. They are typically small and shallow. The force of this grit-filled air blowing against the rock-face gouges out softer types of rock, especially sandstone, to carve caverns. One example is the **Wind Caves of the Anza-Borrego Desert State Park**.

Many caves found around the world are called "wind caves", but are not true wind caves. Most of these are solution caves, but, they are called wind caves because of the strong air currents that blow through them.

One example of a solution cave system that is called a "wind cave" is a maze of caverns and passageways in the United States, called the **Wind Cave in South Dakota**. The Wind Cave's first tunnels began to form about 320 million years ago when mildly acidic rainwater began to flow into the area. Over millions and millions of years, the Wind Cave grew into a maze of over 225 kilometres (140 miles) of caverns and tunnels with a strong air current that sometimes blows into the cave and other times out of the cave – this is called “cave breathing”!

Amazing Cave Formations

Within caves, especially solution caves, minerals can form the most amazing shapes.

Speleothems

Speleothems (pronounced: spee-lee-OH-thems) are wondrous formations found in caves. Thinking about how an icicle is formed will give you an idea of how a *speleothem* can form. Snow falls. As the sun begins to melt the snow, the water drips over the edge of the roof. When temperatures are cold at night, the water freezes into an icicle. The next day, the sun melts more snow. It drips over the edge of the roof, and down the icicle. When temperatures again drop at night, another layer of water freezes on the icicle, making it thicker. Over time, this layering makes the icicle long and thick.

The basic process that forms *speleothems* in caves begins in the same way as the process that forms solution caves: as rainwater encounters plants and soil, rich in the gas carbon dioxide, some of the gas is absorbed. The carbon dioxide turns the water into a weak *acid*, in this example, carbonic acid. The acidic water begins to seep through cracks in the limestone. This *acid* has the power to dissolve a mineral called calcite, which is found in limestone. It carries the calcite in the water. When the acidic water enters the air-filled cavern, the carbon dioxide gas is released from the water into the air. Without the carbon dioxide, the water cannot hold the limestone mineral calcite and this, combined with evaporation, causes small particles of the calcite to begin to layer, just as freezing water layers to create an icicle. Over time, the layers of calcite thicken. This process forms fantastic *speleothems*.

Speleothems appear in many beautiful colours depending on their mineral and *acid* composition – ranging from pure white to yellow, orange, red, brown and even black. The shapes of these beautiful and unique creations are fashioned by how the water enters and moves in the cave. For instance, is it flowing, splashing, dripping, seeping, standing or ponding when it becomes solid?

Here are some examples (Remember to find photographs on the Internet of these amazing shapes):

Stalactites & Stalagmites

Stalactites (pronounced: sta-LAK-tites) and **stalagmites** (pronounced: sta-LAG-mites) are common types of "dripstone" *speleothems.* As the name suggests, "dripstone" *speleothems* are made by "dripping" water. When the acidic water, rich with minerals, forms a drip on the roof of the cave, the air in the cavern causes the carbon dioxide/gas to be released from the water. Without the gas, the water can no longer hold a mineral, such as calcite. This, combined with evaporation, causes small particles of the calcite to be left at the circular base of the drip. These rings of calcite crystals form layers, one on top of the other, over thousands of years, as the water drips, until an icicle-shaped stalactite is formed. If instead the water does not release the carbon dioxide until the drip hits the cave floor, a stalagmite can form, rising up from the floor of the cave.

Stalactites hang from the cave ceiling. They are typically long and come to a point. They are shaped like the icicles that hang from the roof in winter. Stalagmites grow up from the cave floor. They are shaped like cones: think of the shape of orange traffic cones. How to remember the difference between a stalactite and a stalagmite? A "stalactite" is holding on "tight" to the cave ceiling. A "stalagmite", just "might" trip you as it grows up from the cave floor!

Stalactites and stalagmites can take thousands and thousands of years to form. Some researchers estimate that stalactites only grow between 0.6 centimetres (1/4 inch) and 2.5 centimetres (one inch) every 100 years. Brazil's **Gruta do Janelão Caves stalactite** is thought to be one of the oldest and longest in the world, longer than three average school buses lined up. If we do the math, this 28-metre (92 foot) free-hanging stalactite, must be at least 110,000 years old, perhaps over 440,000 years old!

Speleothems are often named for animals or objects that have a similar appearance:

Columns

Cave columns are formed when a stalactite grows far down towards the cave floor at the same time as a stalagmite grows far up, so that the two meet and join to become one floor-to-ceiling pillar or column.

Soda Straws

Cave soda straws are a common type of *speleothem*, formed by the very slow dripping of a water solution rich in calcite. They are long, thin and hollow like the objects they resemble: a soda straw.

Hells Bells

Cave Hells Bells are an unusual type of *speleothem* that can measure almost 2 metres (7 feet) in length. They are found underwater in the **El Zapote Cave** in Mexico. These *speleothems* are bell or trumpet-shaped, hollow and become wider at the bottom, just like bells. Interestingly, they form underwater in an area of freshwater directly above saltwater. Research continues, but it is thought that algae and bacteria likely play a role in how these are formed.

Drapery

Cave drapery are wavy sheets of mineral deposits. They are called cave curtains or drapery because of their appearance. Drapery form on cave walls and floors as water flows rather than drips. When the drapery have striped colour variations or layers, alternating between beige and brown, they are often called, **cave bacon**!

Stegamites

Stegamites that have been discovered so far are black in colour. They form a vertical ridge on the cave floor. Think of a **stegosaurus** and the pointed plates standing up on that dinosaur's back! *Speleologists* theorize that stegamites are formed when the acidic water bubbles up through a crack in the cave floor, leaving a calcite deposit.

Popcorn

You can probably guess what **cave popcorn** looks like? It can appear as little balls or can be chalky white and shaped like movie theatre popcorn! These little knobs are thought to be formed when mineral-rich water drips so slowly that air currents in the cave evaporate the water while it is still a round droplet, leaving calcite in clusters of popcorn-shaped nodules. It is sometimes called **cave coral**.

Frostwork

Cave frostwork is delicate, spikey growths of minerals. They are needle-like with a look similar to **rime frost**. They often grow on top of cave popcorn so *speleologists* suspect evaporation plays a role in the formation of frostwork too.

Dogtooth Spar

Cave dogtooth spar is made of calcite crystals, usually a grouping shaped like the sharp teeth of dogs – or even sharks! They form around the edges of small holes in limestone caves, often giving the appearance of jaws open, ready to bite!

Cave Pearls

Just as pearls form in oysters, so **cave pearls** begin with bits of grit or grains of sand. As water spins the grains, they become coated evenly with calcite. And then, as the water runs over the clustered pearls, it polishes them to a glossy white finish.

Helictite Bushes

Helictite bushes are mineral growths that often look like weird, gnarled trees or bushes from another planet! *Speleologists* still have many questions about these forms. It is believed that they begin to grow as stalactites, but then, for reasons unknown, they begin to twist or to branch out in unusual shapes.

Why they grow like this is a mystery? Theories range from air currents effecting the direction of mineral growth to the action of bacteria in the growth process.

Just like the words stalac-tite and stalag-mite, you may see the word helictites (pronounced: hell-ick-tites) used when these bushes are holding on “tight” to the cave walls or ceiling, and heligmites (pronounced: hell-ig-mites) when they are found on the cave floor.

Crystals

Just imagine exploring deep in the earth, almost 300 metres (1,000 feet) beneath a mountain. You discover a cave of giant crystals glittering with a rare beauty.

The **Cave of Crystals** in Mexico was discovered in the year 2000. This horseshoe-shaped cave sparkles with milky-white and clear gypsum beams shaped like the trunks of massive trees. The largest are about 11 metres long (36 feet) and almost 1 metre thick (3.2 feet). These poles jut out at odd angles turning the cavern into a web of crisscrossing crystal lines.

It is thought that millions of years ago – perhaps 26 million – *magma* forced its way through cracks in the rock heading up toward the surface and finally creating a mountain above this cave. It left crevices in the limestone below. Mineral-rich water flowed into these gaps carving caverns through soft rock and creating an environment for these giant crystals to grow. One researcher compared the cave and its fantastic array of huge crystals to **Superman's Fortress of Solitude**!

These amazing cave formations and others can be found around the world in cave systems like the **Carlsbad Caverns** in the United States, the **Langun-Gobingob Cave System** in the Philippines, the **Škocjan Cave** in Slovenia or the **Reed Flute Cav**e in China.

Cave Food Chain

A *food chain* is one way to describe how a community of living things is linked. It explains the order in which these *organisms* depend on each other for sources of nourishment. To survive, living creatures need energy that we call food. But where could food for cave creatures come from?

Outside the cave, the earth is full of plants that provide food for many animals. But plants cannot grow without light. Instead, plant materials may become a source of food in caves when they are washed in by heavy rains. This flooding carries seeds, leaves and stems into caverns.

Another source of food comes from animals that leave the cave to hunt for their dinner then bring it back to the cavern to eat it. These animals leave scraps and bones on the cave floor.

Animals, like bats, can supply other sources of nourishment. For example, animal droppings like *guano* – yes, that's bat poop! – litter the cave floor. Bacteria and fungi feast on *guano* (pronounced: gwaa-no).

As well, creatures in the cave eat each other! For example, some cave creatures are flesh-eating, which means, they may eat bats or other small creatures they can catch or that fall to the cavern floor.

Here's an example of how the food chain in caves can work. Let's start at the bottom – the cave floor – and then climb the *food chain* ladder:

On the cave floor, on the first rung of the *food chain* ladder, we find *microorganisms* that either make their own food, or feed on what lies on the floor of the cave and in shallow cave pools. *Microorganisms* like fungi, algae, viruses and bacteria, feed on plant and animal material –

remember the bat poop! Then tiny critters – although they are larger than *microorganisms* – feed on the fungi, algae, viruses and bacteria. Examples of these are ghostly white or *translucent* **cave millipedes** and small crustaceans, such as the **Skeleton Shrimp**. These tiny creatures become meals for larger animals that near the top of the food chain, such as **cave spiders**, **cave salamanders**, and **cavefish**.

An example of an animal at the top of the cave *food chain* is the **Amazonian giant centipede**. It can grow up to 30 centimetres (12 inches) in length – as long as a school ruler – and it is full of *venom*. The centipede's front legs, called *forcipules*, have adapted to look like pincers – or powerful fangs! – and are used to inject *venom* into its *prey*. This animal is not considered a danger to people, but it does attack creatures ranging from insects to rats to tarantula spiders! And it even hunts bats.

To trap a bat, the centipede scuttles up the cave wall and holds on to the rock ceiling with the sharp claws on its back legs. The creature swings its body to grab a bat with its front legs as the bat flies by or naps on the wall. The giant centipede then stabs its *forcipules* into the bat to inject its *venom*, much like a wasp with its stinger. This centipede is top of the *food chain* in caves such as Venezuela's **Cueva del Guano** – which means Guano Cave.

Awesome Cave Creatures

When scientists talk about "kingdoms", they are referring to groups or classifications of living things. They may identify as many as seven different kingdoms, such as the Plant Kingdom or the Animal Kingdom.

The Animal Kingdom contains over a million species of animals, including humans! Some members of this kingdom are mammals (like bats), birds, reptiles (like snakes), amphibians (like frogs), insects (like crickets), fish, and arachnids (like spiders).

Spiders! What do people have in common with spiders? The *organisms* in the Animal Kingdom have many differences, but they have similarities too. It is these important things that they have in common that make them all part of the same kingdom. As examples, the bodies of all of us animals, from spiders to humans, contain more than one cell; almost all animals can move around during some of their life stages; and animals eat plants or other animals because their bodies cannot produce the food they need to survive.

A cave may be home to many members of the Animal Kingdom from large ones like bears, to small ones like ants.

Here are just a few of the creatures from the Animal Kingdom that spend time in caves:

Mammals

Mammals are a class or type of creature in the Animal Kingdom. Many of the animals you know are mammals, including humans. All mammals have some similarities.

For example, female mammals produce milk to feed their young. Mammals have backbones. They breathe air. And they have fur or hair. Mammals are warm-blooded, which means their bodies stay at about the same comfortable temperature even if it is very hot or very cold outside.

Here are some examples of mammals that live in caves:

Bears

Some mammals use caves as shelter or even temporary homes. They may give birth to their babies in the cave, escape the weather, or even hibernate. These animals include some species of bears, like the **grizzly bears** that make their winter dens in the **Caves of Bear Cave Mountain** in Canada. Other examples are wolves; raccoons; and large cats, like cougars.

Some prehistoric animals that sheltered in caves have become *extinct*, such as **cave hyenas**, **cave wolves, cave leopards** and **cave bears**.

The giant **cave bear** was much larger than a **grizzly bear**. It was 3.6 metres tall (12 feet) when standing upright on its hind legs. We first learned about this creature when skeletons and paintings on cave walls were discovered. Then, in August of 2020, the body of an ancient cave bear was found preserved when Arctic permafrost began to melt. Cave bears roamed the prehistoric world alongside **woolly mammoths**, **sabre-toothed cats** and the first humans! These bears became *extinct* about 24,000 years ago with the onset of an Ice Age. When a type of animal becomes *extinct*, there are no more of its kind alive on the Earth. Its species has died out and is gone forever.

Bats

These furry mammals have four legs and a thin skin that stretches between their long fingers to create wings. Most bats are *nocturnal* animals, which mean they are active at night and sleep during the day. Caves offer them a dark home where they can safely sleep.

Bats *roost* in caves by hanging from the ceiling. Their claws, called talons, allow them to hold on to the ceiling of the cave. It may seem strange to imagine them hanging around, but the roof of the cave is perfect for flight: bats cannot take off with a run along the ground the way birds are able to because their wings are not strong enough and their back legs are not built for running. Instead, bats climb to the cave ceiling, and then fall into flight.

The **Mexican free-tailed bat** is a cave bat. The colony in **Braken Cave** in the United States is estimated to have 20 million members! In the spring, these bats fly to the cave all the way from Mexico, Central and South America to give birth to their pups. They live there until October. Every night, between 6 p.m. and 8 p.m. a cloud of bats as huge as 48 kilometres long (30 miles) and 32 kilometres wide (20 miles) soars out into the night to hunt insects.

How do millions of bats stream out of the dark cave without causing a traffic jam or crashing into the rock walls – or each other? The answer is that these bats use *echolocation* to recognize what is around them. They use it to detect danger and dinner. Here's how it works: as the bats fly, they chirp, sending out sound waves that bounce off objects and back to their ears. These echoes outline an object for the bat and give the bat information such as the object's size, shape and location. With *echolocation*, bats can avoid objects like the cave walls and, as well, can zero in on the location of insects to eat.

Birds

Birds are a class or type of creature in the Animal Kingdom. As you probably know, they are animals with feathers and wings, even though some birds cannot fly. They are usually born from eggs with hard shells. Birds have backbones. They are warm-blooded, which means they are able to keep their bodies at about the same comfortable temperature even if it is very hot or very cold outside.

Here are some examples of birds that live in caves:

Oilbirds

Oilbirds, like those found in the **Guácharo Cave** in Venezuela, are *nocturnal* like bats, and, like bats, they fly at night using *echolocation*. They sleep in the entrance zone of the cave during the day but leave at night to feed on fresh fruit. The seeds and droppings that they leave on the cave floor contribute to the *food chain* in the same way as bat *guano*.

White-Collared Swifts

White-collared Swifts are found in the **Cave of Swallows** in Mexico. They nest in the cave, but leave to find food. These birds shape their nests from mud and bat *guano*, then line them with grass, bark and plant material. When the nests fall apart, they contribute to the *organic matter* on the cave floor.

At 512 metres (1,680 feet) deep, the Cave of Swallows plunges down into the ground about as far as the **CN Tower** in Canada rises into the sky! The swifts sleep in the cave at night, but each morning, the birds leave the cave to hunt. They begin to fly in a circle, swirling upward and around a cone-shaped path, like a great tornado. It is estimated that a flurry of about two million birds rush out of the cave shaft just after sunrise!

Reptiles

Reptiles, such as **lizards** and **turtles**, are a class or type of creature in the Animal Kingdom. How to recognize a reptile? Most reptiles hatch from eggs that have soft, leathery shells. Reptiles have scales, like a snake, or bony plates, like the shell of a tortoise. Almost all reptiles are cold-blooded, which means that their bodies stay at about the same temperature as the air and water around them.

Here are some examples of reptiles that live in caves:

Snakes

Cave snakes like the **Puerto Rican Boa** and **Ridley's Racer** are not venomous. Instead of poisoning their *prey*, they constrict, which means that they coil their bodies around their dinner tightly, squeezing it until the blood no longer flows back and forth to the heart, causing the heart to quickly stop beating. When the animal is dead, boas swallow it whole.

Cave snakes usually live towards the front of the cavern in the entrance zone or perhaps the twilight zone. This allows them to leave the cave for food or to eat small animals that venture into the cave like rats or lizards. Bats are a favourite food for some cave snakes.

In the **Cave of the Hanging Snakes** in Mexico, **Yellow-red Rat Snakes** slither into cracks and crevices high in the walls and in the ceiling to live. They cannot see or hear but they sense vibrations. At dusk and dawn, as the bats exit or enter the cave, these boas dangle out of their nests, and then strike when they sense their *prey*, catching the bats in their mouths as the mammals fly past.

Crocodiles

In Gabon in Africa, a type of crocodile has been found that appears to spend its adult life inside the cave system. Some crocodiles will make caves a temporary shelter, especially during dry, hot times. But these **orange dwarf crocodiles** are the first to be discovered making the cave a permanent home. Scientists are just beginning to learn about these cave-dwelling crocs. There are many questions to be answered.

Researchers have suggested theories about how the crocodiles became cave-dwellers: some *speleologists* believe that the entrances to the **Abanda Caves** became filled with earth, trapping the reptiles underground several thousand years ago. Today, even where there are some holes, the adults seem to be too large to leave.

Researchers suspect they are evolving into a new species. Like so many creatures that spend their lives in pitch-black caves, these crocodiles are nearly blind. And while crocodiles outside the cave eat fish, shrimps, crabs and frogs, these crocs survive on cave insects, like crickets, and a lot of bats.

Why are they orange? That is a question still to be answered. The first theory proposed was that they were changing colour because they eat orange-coloured bats. But, some researchers now suspect there is a different reason: they believe that the crocs are losing *pigment* (colour) because of the lack of light, but that, instead of turning white, they are turning orange because the cave water they soak in is full of bat *guano*. Bat *guano* contains a compound that can colour the skin orange.

Amphibians

Amphibians (pronounced: am-fib-EEE-ans), such as frogs or caecilians (pronounced: see-sil-ee-ans) like the **ringed caecilian**, are a type of creature in the Animal Kingdom.

How to recognize an amphibian? Amphibians live the first stage of their lives in water; the last part on land. Since most amphibians breathe through lungs and/or through their skin, they must keep their skin moist to be able to absorb oxygen. They have backbones and are cold-blooded, which means that if it becomes cold where they are lying, they must move to a warmer place because their body temperature will match the surroundings. For instance, a toad may move from the shade to the sun to warm itself. Amphibians will eat just about anything they can fit in their mouths! Depending on the size and type of amphibian, it may eat insects, spiders, worms, mice, birds, fish or even bats. Some species spend their whole lives in caves.

Here are some examples of cave-dwelling amphibians:

Cave Frogs

Frogs are amphibians that you have probably seen. There are many types of frogs that shelter or spend some of their life cycles in caves, and a few types that make the entrance zone of caves a more permanent home. For instance, **cave-dwelling frogs** are found in the **Mimbi Caves** in Australia.

Salamanders

Salamanders are amphibians that have the smooth skin of a frog, but have a long tail like a lizard. They live in places that are cool and moist. Some salamanders have lungs; others have gills like a fish. Some breathe through their skin and through tissue inside their mouths and throats. **Cave-dwelling salamanders** may have dark skin or pale skin with very small eyes depending on which zone in the cave they inhabit.

One type is the **Gorgan Mountain Salamander**. It is a rare, critically-endangered species found in Iran's **Shir-Abad Cave.**

Another type is the **Olm salamander**. The Olm lives deep inside the cave, in the dark zone. It spends almost its whole life in cave pools. It even eats and sleeps in the water. This salamander has a snake-like body that grows up to 30 centimetres (12 inches) long. Since its skin lacks *pigment*, it is pinkish-white. The Olm has adapted to complete darkness. For example, its hearing and sense of smell are strong, but since there is no light, it has no need for eyesight. It is believed that an Olm may live for 100 years! This blind amphibian swallows bugs, snails and crabs whole, and because the food supply is limited in caves, its body has learned to survive without eating for more than ten years!

Insects

Insects are a group of creatures in the Animal Kingdom. There are many types of insects that live in caves. You may see **flies**, **fleas** and **cockroaches,** just to name three.

Almost all adult insects have six jointed legs and bodies that are divided into three sections. Most insects have two *antennae* or *feelers*. They are cold-blooded. And they do not have backbones, but their skeletons are on the outside of their bodies, like armour!

Here are some examples of insects that live in caves:

Cave Beetles

Cave beetles are thought to have evolved from above-ground beetles, like the ones you see in your flowerbed. But, your backyard-beetles have plump bodies and short legs. Cave beetles that have changed to adapt to the dark zone of caves often have no eyes, no wings and no colour (*pigment*). Typically, their *antennae* are extremely long, letting them use touch and vibrations, instead of sight, to understand their lightless surroundings. Usually, their six legs are long, like a spider's legs while their bodies are smaller and often bulging in the back section, perhaps storing fat for food because it can be difficult to find dinner in the dark zone of caves.

Cave beetles often remind people of spiders or ants more than garden-variety beetles because of their appearance. The **Narrow-necked Blind Cave Beetle** is a beetle with no eyes, wings or *pigment* (colour). It was first found in Slovenia's **Postojna Cave** system.

Cave Crickets

Cave crickets are another insect found in caverns. Some species live near the entrance to the cave so that they can sneak out at night to find additional food. Others live deep in the cave system. Cave crickets have adapted to life underground: most are active at night (*nocturnal*) and cannot fly. Cave crickets usually have long *antennae* that may be four times the length of their bodies!

One type of cave cricket that lives in New Zealand, in spaces such as the **Aranui Cave**, is the **Cave Wētā**. They have large back legs, like the common grasshopper, and long *antennae*. They cannot fly, but some can jump up to 3 metres (10 feet) in a single leap!

Wētā are big insects: they have small bodies but with their long *antennae* and six very long legs, they can measure up to 40 centimetres (16 inches) from the tip of their *feelers* to the end of their hind legs! They tend to live in the twilight zone of the cave, but may venture out at night to hunt for food. Although Cave Wētā cannot hear, they can sense ground vibrations through pads on their feet, letting them know when *prey* or *predator* is near. They have hard, armoured bodies. Their bodies, especially their *feelers*, are covered with special hairs that sway through the air picking up smells and sensing what is around them even in the darkness of the cave.

Cave Gnats

Elsewhere in New Zealand, in caverns such as the **Ruakuri Cave**, a magical sight greets *spelunkers* and *speleologists* – *spelunkers* are people who explore caves as a hobby.: on the cave ceiling, it appears as if a galaxy of stars is twinkling. But the thousands of sparkling points of light are actually insects called **Waitomo Glowworms**.

These creatures are not worms even though they look like them. They are the larvae of gnats, small flying insects about the size of the common mosquito. Larvae are a stage in the life cycle that some animals, such as most insects and amphibians, go through. Often larvae look a lot like a caterpillar, although not always – think tadpole.

The Waitomo Glowworms begin life as teeny, round eggs that have been laid on the rock face. The eggs hatch into larvae that have no *pigment* (colour) and look like worms. The larvae hang from the cave roof. There they make tubes out of *mucus* – yes, think "snot" – to lie in, like a gooey hammock! Then they use their mouths to make silk threads that they hang down from the cave ceiling and coat in thick, sticky *mucus* (pronounced: myoo-kus). The threads are like fishing lines to catch their dinner!

A chemical in the larvae body causes them to emit a blue-green glow to attract food to the sticky strands. Insects are fooled into thinking the lights from the glowworms are stars in the sky. When the insects fly upward, they are trapped, stuck in the *mucus*. The larvae reach down or haul up the threads so that they can feed on what has been caught. It takes a few minutes, but if disturbed, the larvae can turn off their lights!

Arachnids

Arachnids (pronounced: ah-RACK-nids), such as spiders and scorpions, are a group of creatures in the Animal Kingdom. Like insects, they are cold-blooded and they do not have backbones, but their skeletons are on the outside of their bodies. Unlike insects, most adult arachnids have eight legs and usually have only two sections to their bodies. They do not have *antennae*.

Some of the creepiest cave creatures are arachnids, probably because of the way they have adapted to the dark zone gives them a truly alien appearance.

Here are some examples of arachnids that live in caves:

Cave Spiders

The **Cave Robber Spider** is found in the **Oregon Caves** in the United States. This type of spider was only captured for the first time in 2010 so scientists are still studying it to get answers to questions such as what it eats.

Although researchers do not know what it likes for dinner, they believe they know how it catches its *prey*: these arachnids have sharp and jagged, hook-like claws. Scientists think that they weave strands of silk into their webs on the cave ceiling and then dangle from those strands. When *prey* comes near, the spider uses its razor-sharp claws to attack, inject *venom* into its victim and then bring it home for dinner.

Cave Pseudoscorpions

The National Speleological Society's choice for USA Cave Animal of the Year for 2020 was an arachnid called the **Great Basin Cave Pseudoscorpion**, found in the **Lehman Caves** in the United States. This creature is a distant relative to other cave arachnids like **giant Laotian harvestmen**, also known as a Daddy Longlegs; **Texas cave scorpions;** and **cave mites**.

The prefix "pseudo" means "false" so pseudoscorpions (pronounced: sudo-scor-pee-ons) are not true scorpions. These creatures do have claw-like arms like scorpions, but they do not have the scorpion's stinging tail. Instead, pseudoscorpions inject paralyzing venom into their *prey* using special teeth.

The Great Basin Cave Pseudoscorpion eats small insects and mites. But arachnids do not take bites of their food and chew it before swallowing as we do. Instead, arachnids enjoy a liquid lunch! Some arachnids, like ticks, suck blood. But, others have a more gruesome way of preparing their meals: the Great Basin Cave Pseudoscorpion grabs hold of its *prey* with its pincers. It then injects venom into the *insect* to paralyze it. The juices in this arachnid's stomach and digestive tract are able to break down its *prey's* body, turning it into a liquid. So, the pseudoscorpion vomits these juices onto the insect to turn it into a wet goop. As the animal that it has trapped for dinner "melts", the arachnid sucks up the liquid meal. Yuck!

Fish

Fish are a class of creature in the Animal Kingdom. As you know, they live in the water. But did you know that they "breathe" water! Fish take water in through their gills and remove the oxygen (air) from it. They are animals with backbones, scales and fins. Fish are cold-blooded, which means that their bodies stay at about the same temperature as the water around them. Fish can be different shapes, sizes and colours, such as the beautiful **Mandarinfish** or the huge **basking shark**.

Cavefish have adapted to life in underground lakes and rivers. They often have to survive in the dark zone of the cave with little food. Because there is no light, most cannot see and have no colour (*pigment*). They may have large fins but no scales. The ability to sense vibrations helps them find *prey* and avoid obstacles. Many have a strong sense of smell and taste.

Here are some examples of fish that live in caves:

Mexican Blind Cavefish

The **Mexican blind cavefish** is a cave form of tetra fish, a freshwater species of fish originally from Africa, Central America and South America. The Mexican blind cavefish has been found in the **Chica Cave** in Mexico.

The cavefish form of the testra fish is thought to have appeared about 20,000 years ago. Scientists have discovered that these sightless cavefish have a unique way of avoiding obstacles, a way similar to *echolocation*: they suck water into their mouths, and then release it through their gills. This creates currents in the water around them. Instead of sound waves, like in *echolocation*, these water waves bounce off objects like rocks or pond walls. Sensors on their bodies detect the water movements and allow the fish to recognize its surroundings.

Waterfall Climbing Cavefish

A fascinating fish adapted to survival in the dark zone is the blind **waterfall-climbing cavefish**, native to Thailand and found in caves, such as the **Tham Mae Lana**. The fins on these cavefish stick so fiercely to rock walls that they can even climb when fast-running water, like a waterfall, is pushing against them. As they grab hold with their fins, they wiggle their hips, moving much like a salamander, to climb the rock face. Researchers suspect that as it climbs, it finds algae and very small *organisms* and *microorganisms* that it can eat.

Every year, *spelunkers* and *speleologists* discover new caves and cave creatures. In 2019, a new **cave-dwelling Dipluran** was found in **Kiku Pot Cave** in Canada. In 2020, the poisonous **King of the Cave Centipede** was discovered in the **Movile Cave** in Romania. And in 2021, in the **Valencia Caves**, it was reported that a new **cave-dwelling Kinnaridae,** named Fairy of the Valencia Forest, was identified. Imagine touring a cave that no modern human has ever before entered or discovering a new species!

Have humans ever lived in caves?

Some prehistoric people sheltered in caves. We know this, in part, because they created drawings on the walls and ceilings. But, how much time did they spend underground? Did some humans live in caves all year long? Perhaps they only stayed in caves during the winter or rainy season? Maybe they entered caves to perform rituals or to use them for storage, but did not use them as permanent homes? These are the types of questions that researchers are trying to answer.

Ancient artists created their drawings with colours they obtained from natural substances. They used these *pigments* to give paint colour. *Pigments* came from minerals, plants, animals, soils and rocks. They mixed these substances with fat to create paint. Prehistoric artists could produce colours that ranged from shades of yellow to orange to red to brown to deep purple and black depending on the mineral and mixture used to paint. For example, black could be made with ash or charcoal from fires. Blue and green do not seem to be used by ancient humans in their artwork. It may be because rocks containing blue minerals to make into paint were difficult to locate? And perhaps green was not used because it was difficult to see on the damp, mossy walls of the cave? Scientists are still searching for answers to so many questions.

The paint was applied to the rock walls and roofs in many ways. We know that brushes could be made from grasses, feathers or even horse hair. Burned sticks offered charcoal tips for sketches, much like a pencil. Hollow bones could be used to blow colour onto the rock wall. Paint could be sponged on with moss. And hands could smear paint on large areas quickly.

Imagine sitting at a roaring fire, flames flickering off the cave walls. As you paint, the images come alive with colour. You may draw some of the animals you hunt. Or maybe the animals that hunt you like **cave lions** and **cave hyenas**!

Here are some examples of cave art:

In the **Blombos Cave** in South Africa, crossed lines resembling the symbol for "number sign" or "hashtag" (#) have been dated at about 73,000 years old. In 2018, one scientific study suggested that the sketch of a ladder found in a cave in Spain was painted 64,800 years ago. Others, such as a hand print – the artist would have put colour on his hand and pressed it against the rock – were found in **La Pasiega Cave**, **Maltravieso Cave** and **Ardales Cave**. Recently, a drawing of a **babirusa**, found on the rock wall of the **Leang Tedongnge Cave** in Indonesia, was dated at about 45,500 years ago – babirusas are hairless wild pigs with huge curved tusks that resemble a deer's antlers so they are sometimes called pig-deer.

Around 31,000 years ago, artists at the **Chauvet Cave** in France were painting grazing animals like horses, but also the *predators* that threatened them, such as cave bears, hyenas, lions and **Ice Age leopards**. In the **Lascaux Cave** system, also in France, over 600 multi-colour paintings and charcoal drawings decorate the ceilings and walls. The art in this cave and others in France, such as the **Rouffignac Cave**, were painted at least 13,000 years ago. They feature images such as **woolly rhinoceroses** and mammoths, bison, deer, fish, horses, birds and even an *extinct* species of wild cattle called **aurochs** (pronounced: or-rocks).

The **Magura Cave** in Bulgaria has more than 700 drawings dated to between 4,000 to 8,000 years ago. Bat *guano* – yuck! – was used to create these pictures of dancing people and hunting parties following a large variety of animals.

Who were these early humans?

N*eanderthals* (pronounced: nee-AN-der-thals) are one species of early human. Think of them as our cousins! They are thought to have become extinct about 35,000 years ago, perhaps as recently as 28,000 years ago according to 2006 dating from **Gorham's Cave** in Gibraltar.

We have more clues to help us understand ***Neanderthals*** than other *extinct* groups of early humans because we have found evidence, such as *fossils* and *artifacts*. *Artifacts* are objects, such as stone tools or jewellery, which give us information about a culture or society. *Fossils* are the preserved remains or impressions of *organisms* that lived at least 10,000 years ago, such as plants and animals from long ago. Examples of fossils are hardened footprint impressions, bones and teeth.

Fossils are nature's record of prehistoric life. A fossil can be made when the plant or animal is buried and then the minerals from the dirt and debris that covered it turn the remains into a stone copy of the original. This stone copy is called a fossil.

One exciting example of an ancient skeleton is called **Altamura Man**, found in the **Lamalunga Cave** in Italy. This *Neanderthal* lived between 128,000 and 187,000 years ago. Scientists were able to use his DNA to gain a better understanding of his people and how they lived.

Here is some of what is believed about *Neanderthals*:

Neanderthal adults were shorter and more muscular than today's humans. Adult males were not typically taller than 1.75 metres tall (5'7"). Scientists think that

Neanderthals were stocky, meaning they had broad shoulders but short legs and arms. Their faces had large noses, a thick ridge where their eyebrows sat, a weak chin and big teeth.

Scientists believe that *Neanderthals* did very little farming, if any. Instead, they hunted, fished and gathered food like berries and nuts that grew wild. They knew which plants could be used as medicine. Braided plant fibres could be used as yarn and string. They were able to weave their own simple clothes and also worked with feathers and fur. They knew how to turn animal hides to leather.

Although *Neanderthals* spent time in caves, they also knew how to build outdoor shelters. But, no lumber and nails for them! Imagine a hut built with woolly mammoth bones lashed together with leather strips. *Neanderthals* were skilled at making cooking vessels, tools and weapons. They made them out of bone, wood and stone, such as flint.

Flint is a hard rock used for spear tips and tools, like hand axes called bifaces. *Neanderthals* made fire by striking the flint biface against a mineral like pyrite to make sparks. The sparks would light materials such as dry grasses and twigs. With fire, these early humans could heat birch bark to turn it into a tar that could be used for many purposes, including glue.

Some evidence suggests *Neanderthals* travelled by water, which would mean they built rafts or watercrafts. In **Figueira Brava Cave** in Portugal, scientists found evidence that *Neanderthals* ate seafood, gathering mussels and crabs, and catching fish and even sharks! As well, these early humans followed migrating animals to hunt them for their meat.

Neanderthals used every part of the animals they hunted: the meat, fat and organs were rich sources of nutrients so were eaten; the stomach and intestines could be used as water containers; sinew, the ropey tissue that holds body parts like muscle and bone together, can make strong cords. These cords could be used to attach a flint axe blade to a wooden handle or to string a bow or to hold wood together to build a shelter. The skin of animals could be dried or tanned to serve many purposes such as clothing or to make bowls and buckets. Bones made excellent utensils, tools and weapons. Feathers and quills could be used for decoration or even as paint brushes.

Modern Humans and Caves

Researchers have discovered that modern humans, like you and me, have a small amount of *Neanderthal* in our genes. A modern human is called a ***Homo sapiens*** (pronounced: ho-mow-say-pee-en). It makes sense that during the time *Neanderthals* and modern humans lived on this Earth together that some children would be born with a parent from each species of human. Just as your genes may give you the eye colour of one parent and the hair colour of the other parent, so a child with both a *Neanderthal* and a *Homo sapiens* parent would share characteristics with his or her parents. And these traits would be handed down to their child's children and on and on through the generations.

Modern humans began to farm, rather than just hunting, fishing and gathering like *Neanderthals.* Because agriculture and domesticated animals were so important to their survival, they built permanent dwellings to live in while they raised their crops. Most did not make caves their permanent homes. But, cave paintings that were drawn less than 25,000 years ago are thought to be the work of modern humans rather than *Neanderthals.* So, scientists believe that caves were places for human activities, such as religious ceremonies and even to shelter from the sun on a hot day. Because caves are cool, they could be used for some storage too, like your refrigerator!

We only know of a few areas in more modern times where people have made permanent homes in nature's caves. And in most of these caves, modern humans have dug them out to make them larger or renovated to make them more comfortable. A human who lives in a cave is called a *troglodyte* (pronounced: trog-LOW-dite).

There is evidence of *Homo sapiens* cave-dwellers in the American Southwest as early as prehistoric times, often living in caves dug out by the wind. The **Ancestral Puebloans**, an ancient Native American culture, made their homes in caves or built homes in the wind-carved spaces under rock overhangs along canyon walls. At **Mesa Verde** in the state of Colorado, these people built amazing cliff dwellings, beginning with shallow caves that they then enlarged. Often the cave was only deep enough for one or two separate spaces. But the **Cliff Palace**, built over 700 years ago, between 1190 C.E. and 1260 C.E., was carved and constructed to create a palace of about 150 rooms. The rooms were made of sandstone, wood beams and mortar mixed from soil, water and ash.

The **Rock Houses of Kinver** in England are thought by some to have been the inspiration for J.R.R. Tolkien's hobbit homes! You can be found under Kinver Edge, a 250-million-year-old escarpment made of sandstone. Over time, the soft sandstone had been worn away in areas to create shallow caves. **Nanny's Rock** is one of these caves. This five-room cavern is thought to have been home to a woman named Margaret in 1617. Kinver Edge's soft rock made it easy for humans to deepen some of the other natural caves to create rock houses. People added doors and glass windows to the fronts. Some of these were lived in as recently as the 1960s.

The original settlement for what is now the city of Matera in Italy was in two nearby canyons. The caves there have been occupied from prehistoric times to present day. In the last thousand years, up until about the 1950s, the very poor lived in the caves. But today, some of these caves have been renovated to become luxury homes and you can even stay at a five-star cave hotel, **Aquatio Cave Luxury Hotel & Spa**!

In China, Zhongdong is a village of small wood and bamboo huts located within the **Zhong Cave**. About 100 people live year-round in this natural cavern. The area of the cave is about the size of four and a half NFL football fields. Wood fires provide heat, but the people have electricity to run lights and small appliances like washing machines and televisions. For their children, the villagers managed what may have been the world's only school in a cave until it closed in 2011!

Careers Studying Caves

Caves are fascinating. Certainly, a career studying caves would be amazing! There are many careers where caves or what is found in caves are studied. For example, archeologists (pronounced: ar-KEE-ol-oh-jists) study *artifacts* (objects) usually from ancient or prehistoric times to understand what those people were like and how they lived. Where archeologists typically analyze physical evidence, anthropologists (pronounced: an-throw-pawl-OH-jists) are usually most interested in human beings and the societies they live in. Some anthropologists are interested in prehistoric times to discover how humans evolved and how they lived together compared to people today. *Speleologists* try to understand everything they can about cave systems, such as what the cave is made of, how it formed, what it can tell us about climate change and what species of plants and animals live in the cave. *Speleologists* are often experts in many fields of science.

Archeology

Dorothy Garrod was an archeologist who uncovered evidence of early humans in caves. Her first discovery was in 1925 at an excavation site called **Devil's Tower Cave** in Gibraltar. She found a *Neanderthal* skull and other evidence to show that it was used as a shelter by these early people. During her career, she made many discoveries, including finding some of the earliest evidence for when humans tamed and domesticated dogs. She is greatly respected for recognizing archeology as a distinct field of science and for introducing a scientific approach to digging at excavation sites so that all the evidence was carefully measured, recorded and preserved. Today's archeologists are like detectives: they search for clues, gather evidence, study what they find, decipher its meaning and then share what they have discovered with the world. Every discovery is a piece in the puzzle that the archeologist will put together to create a picture of how those people lived.

Anthropology

In 1938, wife and husband team, Mary and Louis Leakey, excavated the **Njoro River Cave** in Kenya. There these anthropologists discovered that the cave was a burial place containing about 80 Stone Age skeletons and grave goods such as pottery, stone bowls, baskets and thousands of beads.

One of most significant discoveries by **Mary Leakey** occurred in Tanzania in 1976 when she uncovered *fossils* of animal prints that were about 3.6 million years old. By 1978, the researchers had been able to recognize that some of these tracks, called the **Laetoli Footprints**, belonged to early humans. They were made when three prehistoric humans, walking on two feet instead of four feet, slugged through wet volcanic ash. Those footprints hardened and were then preserved by additional layers of ash. This was the first true proof of when early humans walked upright instead of on their hands and feet like a **chimpanzees**.

Louis Leakey is also remembered for recruiting three outstanding women to live in the wild with great apes to study them: **Jane Goodall (chimpanzees)**, **Dian Fossey (mountain gorillas)** and **Biruté Galdikas (orangutans)**.

Speleology

Dr. Penelope Boston, is interested in caves not only because she is a *speleologist*, but because she is also a microbiologist (a scientist who studies *microorganisms*), an astrobiologist (pronounced: as-troh-bi-ol-oh-jist) and became Director of the NASA Astrobiology Institute in 2016. Astrobiologists study life in the universe, including what is required for there to be life on other planets. Oh yes, and she speaks Klingon!

What is the connection between caves on Earth and the study of life on other planets? For one thing, scientists have found a type of cave on Earth (lava tube) that also exists on the Moon and on Mars! Could Earth's astronauts survive in lava tubes on Mars? Studying caves and cave creatures may provide an answer.

As well, because the environment in caves is so different from life on the surface of the Earth, caves may teach us about alien planets. For example, the layer of gas (air) that surrounds the Earth is called the *atmosphere* (pronounced: at-moh-sfeer). Earth's *atmosphere* is made up of about 21% oxygen, making it easy for us to breathe. We cannot breathe the air on Mars because the *atmosphere* on that planet is about 95% carbon dioxide with very little oxygen. Like the planet Mars, caves are often low in oxygen and can contain poisonous gases. So, studying the creatures in caverns that can survive in that toxic air may offer us clues of how to breathe in the Martian *atmosphere*.

Dr. Boston is also interested in discovering new life forms in caves. Why? Because these critters have found a way to survive without sunlight or the food sources that we have on the surface of the Earth. These are problems that we may have to overcome if we try to live on Mars! What scientists learn about caves and the creatures that thrive in them may help with the exploration or colonization of our solar system.

When exploring Mexico's **Cave of Crystals**, Dr. Boston and her team discovered *microorganisms* alive in air bubbles within pockets of fluid in the crystals. These *microorganisms* may have been trapped there 50,000 years ago! What can we learn about how they survived?

In another cave in Mexico, called **Cueva de las Sardinas**, Dr. Boston and fellow scientist, Diana Northup, found what looked like a slimy stalactite dripping from the ceiling. Because the slime appeared to be a lot like the *mucus* or "snot" – think your runny nose when you have a cold – she called it a "snot-tite".

Reseachers have learned that snottites are a community of *microorganisms*, specifically bacteria, which produce this goo and then live in it. They do not need sunlight to survive. Unlike plants that rely on the energy from sunlight to make their own food through a process called *photosynthesis* (pronounced: foh-TOH-sin-thuh-sis), *chemotrophs* (pronounced: key-MOH-troefs), like the bacteria in snottites, are *microorganisms* that get the energy they need to survive through chemical reactions. This allows them to live in dark, harsh environments like caves – or perhaps on distant planets!

The **Caves of Mars Project**, funded by the National Aeronautics and Space Administration (NASA) studied, among other things, the potential for manned missions to the **planet Mars** where astronauts would live underground.

With information from scientists, space probes can search for signs of past and present life in alien environments, such as in the caves on the planet Mars. The more we learn about different creatures of our world and how they adapt to their environments, the more this knowledge will help us to understand, not just our world, but others in the universe.

More Mysteries in Caves

Caves are fascinating spaces. Every one of them is unique. There are still so many questions to be answered and so many secrets to be revealed.

Scientists never know what they may find when exploring cave systems. Caves can tell us the story of animals and people long gone from this Earth. They are full of mysteries to be solved!

And amazing creatures live in this underground world ... If you dare, search the Internet for photographs of these alien-looking, otherworldly critters found in caves:

Bayliss Cave Planthoppers
Blind Cave Prawn
Brazilian Blind Characid
Cave Leeches
Cave Waterlouse
European Cave Spider
Giant Cave Cockroach
Humpback Golden Line Barbel
Kaua'i Cave Wolf Spider
Lava Tube Crabs
Madagascar Blind Snake
Peck's Cave Amphipod
Tailless Whip Scorpion
Tumbling Creek Cave Snail
Two-pronged Bristletail
Youngsteadt's Cave Millipede

Cave Crossword

Need help with answers, check the *glossary*!

<u>Down</u>

1. Wearing away of land by wind and water.
2. Hot liquid rock on the earth's surface.
3. When a species is gone forever.
6. Chemical substances that erode materials.
7. It hunts other creatures for food.
9. The envelope of gases around a planet.
10. Where birds/animals regularly rest.
13. Bat and seabird poop!

<u>Across</u>

4. Living things.
5. Active at night and asleep during the day.

8. Substance that gives something colour.
11. Humans who live in caves.
12. A scientist who studies caves.
14. A species of early human.
15. A hollow space carved by nature and fits a person.

Glossary

A *glossary* is an alphabetical list of words with their definitions.

Acids: Chemical substances that can eat away some other materials.

Antennae: Long, thin parts on the heads or front sections of creatures such as insects. Antennae are also called *feelers*.

Artifact: Artifacts are objects found at archeological dig sites and that were made or shaped by humans, such as stone tools, metal weapons and jewellery.

Atmosphere: The envelope of gases around a planet. We also call it air.

Chemotrophs: Organisms that use chemical reactions to obtain the energy/food they need to survive.

Diurnal: Diurnal describes the behaviour of a creature that is active during the day and sleeps at night. The opposite of diurnal is *nocturnal*.

Echolocation: Using sound waves to locate and identify objects. Animals like whales, dolphins, porpoises and bats use echolocation.

Erosion: The wearing away of soil, rock and land by forces of nature like water.

Extinct: When every member of a species of animal is gone from the Earth forever, we say that type of animal has died out or become extinct.

Feelers: Another name for *antennae.*

Food Chain: The order in which living things in a community depend on each other for food. Here is a simple example: Grass is nibbled by a cricket. The cricket is then gobbled up by a frog. The frog is swallowed by a snake. The snake then becomes dinner for a hawk.

Forcipules: The forcipules on a centipede are a pair of legs adapted for use in piercing the skin and injecting *venom* into its *prey.*

Fossils: The preserved remains or impressions of living *organisms* that lived at least 10,000 years ago.

Guano: The excrement or droppings of seabirds and bats.

Glacier: Huge ice masses that form over periods of time. The oldest glacier is thought to be in Antarctica and may have formed as long as one million years ago.

Homo sapiens: A modern human, like you, is called a *Homo sapiens.*

Lava: Intensely hot liquid and semi-liquid rock that breaks through the earth's surface when a volcano erupts. When this liquid rock cools to become solid, we continue to call it "lava", but when the *molten* rock is still within the earth, instead of *lava*, we call it "*magma*".

Magma: Intensely hot liquid and semi-liquid rock below the earth's surface. When the magma breaks through to the surface of the Earth, then scientists call it "*lava*".

Microorganisms: "Micro" means very small; "*organisms*" means living things. So, "microorganisms" are very small living things, such as bacteria.

Molten: Something at such a hot temperature that it has become a liquid.

Mucus: Sometimes called "snot", it is slimy, sticky stuff. One type is what you see running from your nose when you have a cold.

Neanderthals: One species of early human.

Nocturnal: Nocturnal describes the behaviour of a creature that is active at night and sleeps during the day. The opposite of nocturnal is *diurnal*.

Organic Matter: The remains from plants and animals. Examples of organic matter are manure on a farmer's field, guano in a bat cave or leaves on a forest floor.

Organisms: Living things.

Pigment: Substances that give something colour.

Photosynthesis: The process by which green plants use light to make their own food.

Predator: A creature that hunts other critters for food. It captures *prey*.

Prey: A creature that a *predator* hunts for food.

Roost: A place where birds or animals, such as bats, regularly settle to rest.

Speleologists: Scientists who study caves.

Speleology: The scientific study of caves.

Speleothems: Mineral deposits that form in caves.

Spelunker: A person whose hobby or sport is exploring caves.

Translucent: Almost see-through or semi-transparent.

Troglodyte: A human who live in a cave.

Venom: A type of poison used by some animals.

Made in United States
North Haven, CT
20 November 2023

44298494R00036